## About the Author

**Vijay Gurav** is an accomplished industrial engineer with over a decade of experience in designing and optimizing manufacturing and production assembly Line systems. A certified Six Sigma Black Belt, Vijay holds a Master's degree in Industrial Engineering from the University of Texas at Arlington and a Bachelor's degree in Mechanical Engineering from the University of Mumbai.

Throughout his career, Vijay has specialized in crafting innovative solutions for both single-line and mixed-model assembly lines, helping organizations achieve unprecedented efficiency and productivity. His expertise lies in seamlessly integrating advanced technologies—such as artificial intelligence and automation—into traditional manufacturing processes to drive innovation and transform operations.

With a passion for continuous improvement, Vijay has been at the forefront of modernizing manufacturing practices. His dedication to excellence, strategic thinking, and ability to bridge the gap between traditional engineering and emerging technologies has earned him a reputation as a leader in his field.

In this book, Vijay shares insights, strategies, and lessons learned from years of hands-on experience, empowering industrial engineers to unlock their full potential and join the top 1% in the industry.

When he's not innovating in the manufacturing world, Vijay enjoys mentoring aspiring engineers and exploring new developments in technology and operational excellence.

**Connect with Me!**

Feel free to reach out via email at **vijaygurav1@outlook.com** or connect with me on LinkedIn.

I'm always open to discussing new ideas, opportunities, and professional connections!

# Title

1. History of Industrial Engineering ............................................................ 7
2. What is Industrial Engineering ............................................................. 11
3. Production Safety .................................................................................. 17
4. Plant Productivity & Production Cost ................................................. 20
5. Factory Capacity Planning .................................................................... 27
6. Work and Method Studies .................................................................... 32
7. Tools in Industrial Engineering ............................................................ 39
8. Lean Manufacturing ............................................................................... 44
9. ERP Systems ........................................................................................... 52
10. Assembly line Design .......................................................................... 55
11. Bottleneck Analysis ............................................................................. 64
12. Engineering Economics ...................................................................... 67
13. Finance Basics for Industrial Engineers ........................................... 72
14. AI in Manufacturing & Factory Floor ............................................... 76
References .................................................................................................. 80

# 1. History of Industrial Engineering

In the late 18th and early 19th century, during the Industrial revolution there was huge boom in manufacturing process from manual labor to mechanized production. Industrial Engineering, Management science and efficiency, systematic organization started developing dueting this time. There was high growth in demand for products with limited resources & time available to products, thus developing science of factory & industrial engineering concepts.

**Frederick Winslow Taylor,** born in 1856 in Philadelphia, Pennsylvania, was a Mechanical Engineer, and is usually given credit for being Father of Industrial Engineering and Father of scientific management. He was also one of the first Management consultants in USA. In 1911 he wrote The Principles of Scientific Management, which was one of the most influential books at that time. His Main principles were:

1. Collect Data on work elements & develop standardize work.
2. Development and Train workers provide details instruction on best way to do work.
3. Work division between Management & workers so each group does best suited work.
4. Train Managers how to analyze each job.

He also was pioneer in developing Time Study which every Industrial Engineer and production facilities in modern world still use it, a simple use of stopwatch that would recommend a stand method for time taken for a given work element to maximize efficiency, maximum output and minimum fatigue. Lter in future

7

this was combined with Gilbreths methods study, new discipline of 'time and motion study' was born.

**Henrey Laurence Gantt** worked with Fredrick Taylor between 1887 to 1893, created Gantt Chart which significantly helps managers to visually manage large scale projects and scheduling needs. These charts are still used today on large scale projects for planning, scheduling, monitoring and reviewing the work. Famous projects include Hoover Dam & construction of USA interstate highway systems Gantt charts are significantly used.

**Frank Bunker Gilbreth**, Bron in 1868 Fairfield Maine, Served in USA Army during World War I. He developed training procedures to train solders on how to dissemble and reassemble their firearms quickly, even if it's dark or when blindfolded. He was able to capture every person's motion in 17 Basic Motions (Later 18 motions) called as 'Therbligs' which is Gilbert spelled backwards. This eventually became a building block for PMTS (Predetermined Motion Time Study) MTA (Method time Analysis. Instead of stopwatch Gilbreth used a motion picture camera to film work done by person and was able to determine time and work standards also considering motion. Taylor and Gilbreth work together helps to reduce waste and motion at work. His work also developed recommendations for redesigning tools, workstations & thus leading into Human factors engineering and ergonomics.

**Henry Ford**, Bron in 1863 on a farm in greenfield Township Michigan founded the Ford Motor Company. He revolutionized manufacturing by creating the assembly line to produce automobiles. His principles of developing mass production assembly line with smaller task reduced time to build Model T automobile from 700 hours to 1.5 hours. His mass production methods reduced cost and common thus helping everyone's dream to own a car. He significantly improved wages, and reduced work hours by 48 hours a week to 40 hours per week. This principle of mass production system and assembly line today are adopted by every industry and have set foundation for modern manufacturing and factory design.

**Walter G. Holmes** wrote a book on Applied Time and Motion Studies when he was a Time study Engineer at Timken Detroit Axle Company in Detroit Michigan in 1930s. His books describe ways to conduct time studies, use of process flow charts and workflow diagrams. He explained how to level and average time studies, conduct (PMTS) Predetermined Motion Time Systems and use to Therbligs in production.

**Dr. Walter Andrew Shewhart**: Born in 1891, Shewhart received a doctorate in physics from University of California, Berkley worked as an inspection engineer at Western Electric company inspecting final products and discarding defective ones. He started using controls charts in manufacturing processes thus resulted in use of Quality control. Today you will see the Quality control and inspection department in every major company and production facility. He started using statistical methods to control variation in production process and try eliminating it from the root cause of this variation from production process.

**Dr W. Edwards Demings**: Born 1900 in Iowa, he has an electrical engineering degree and master's from Yale University, he began his work with Shewhart during World War II to improve productivity and Quality. In 1950, Demings started using Shewhart theories in Japan to recover production and manufacturing from devastation of World War II. He is also famous for implementing (PDCA) Plant, Do, Check, Act still used widely in Industry. Demming was awarded the National Medal of technology, and his fundamental philosophy is summarized in 14 Principles of Management.

| 1 | Constancy – Purpose | 8 | Break – Silos |
|---|---|---|---|
| 2 | Adopt – Philosophy | 9 | Optimize – Quality |
| 3 | Cease – Dependence | 10 | Eliminate – Slogans |
| 4 | Improve – Processes | 11 | Remove – Quotas |
| 5 | Train – Workforce | 12 | Enhance – Pride |
| 6 | Leadership – Guiding | 13 | Encourage – Learning |
| 7 | Drive – Fearless | 14 | Act – Transformation |

# 2. What is Industrial Engineering

"Industrial Engineering is concerned with the design, improvement, and installation of integrated systems of people, materials, information, equipment, and energy. It draws upon specialized knowledge and skills in the mathematical, physical, and social sciences, together with the principles and methods of engineering analysis and design." – Institute of Industrial and Systems Engineers (IISE)

The main objective of Industrial engineering is to increase productivity by eliminating waste and non-value adding, unproductive operations and improving effective utilization of resources. It focuses on designing, optimizing and improving complex systems, process and organization. It aims to integrate man, material, equipment, energy and information to achieve productivity, efficiency and quality of production and operations.

Industrial Engineering has also expanded into service industries due to techniques such as

1. Operational Research
2. Scheduling and Project Management
3. Human Factors Engineering
4. Value Engineering
5. Systems Analysis
6. Advances in Informational Technology
7. Mathematical and Statistical Tools
8. Reliability and Maintenance
9. Queuing systems in Healthcare & Theme Parks

With the modern world of the internet, it's important to understand for fresh graduate with degree in Industrial Engineering, Lean Engineering or Manufacturing engineering which roles are available to apply on job hunt websites:

**Roles and Job Title for Industrial Engineering: -**

1. Industrial Engineer
2. Method Engineer
3. Process Engineer
4. Quality Engineer
5. Manufacturing Engineer
6. Operations Engineer
7. Innovation Engineer
8. Human Factor Engineer
9. Production Engineer
10. Supply Chain Analyst/Engineer
11. Project or Scheduling Engineer
12. Systems Engineer

More specialized roles which with more experience Industrial Engineer can Apply

1. Automation Engineer
2. Process Automation Analyst
3. Tims study Analyst
4. Industrial Data scientist
5. Sustainability Engineer
6. Operations Research Analyst
7. Cost Estimator
8. Procurement Analyst
9. Buys and Purchasing Analyst
10. Data Analyst (Industry Focus)
11. Six Sigma Black Belt
12. Lean Manufacturing Engineer
13. Continuous Improvement Engineer

## Mid Manager Roles for Industrial Engineer

1. Operations Manager
2. Production Manager
3. Supply Chain Manager
4. Manufacturing Engineering Manager
5. Process Engineering Manager
6. Continuous Improvement Manager
7. Supply Chain Manager
8. Project Manager

## Senior Leadership Roles for Industrial Engineer

1. Director of Operations
2. Director of Supply chain
3. Vice President of Operations
4. Chief Operating Officer (COO)
5. Vice president of Quality
6. Vice President of Continuous Improvement
7. Plant Manager/ General Manager

## Typical Responsibility of Industrial Engineer

1. Standard Work: Develop the simplest way of performing a task or work.
2. Standard Time: Develop Labor standard and performance standard for work.
3. Develop wages and incentive scheme
4. Develop Quality standards, perform audits and analyze defects
5. Estimate cost, performance cost, Return on investment calculations
6. Optimize supply chain, determine Economic order quantity, lot size.
7. Design Assembly line and production systems
8. Design factory layout and production process maps
9. Develop strategic planning with business and operations long term goals
10. Improve and design productivity systems and ways to improve them
11. Waste and non-value-added work reduction methods and training
12. Develop Production planning and schedule optimization
13. Support and Develop Order forecasting, labor and resource needs and forecasting methods.

## Add on Certifications for Industrial Engineers:

1. Six Sigma Certifications by ASQ
2. Lean Six Sigma Certifications by ASQ
3. Certified Quality Engineer by ASQ
4. Certified Supply Chain Professional (CSCP) by APICS

5. Certified in Production and Inventory Management (CPIM) by APICS
6. Project Management Professional (PMP) by PMI
7. Certified Scrum Master (CSM)
8. Certified Manufacturing Engineer by SME
9. FE (Fundamental of Engineering) and PE (Professional Engineer)
10. ISO 9001 Lead Auditor certification

Software is always a tool to make things easier for engineers. There might be many more software available, but I would like to mention a notable few which are good starting points got modern industrial engineers and assembly line design industrial engineers to learn and use in day-to-day applications.

### Software for Industrial Engineers: -

1. MS Excel
2. MS Visio
3. Arena Simulation, SIMUL8, Flexim
4. Cplex (IBM)
5. Gurobi
6. Primavera P6
7. AutoCAD, SolidWorks, Catia, NX CAD
8. SAP, Oracle SCM Cloud, AS400 IBM
9. Minitab, R Programming, Tableau, Power BI
10. Microsoft Dynamics 365, ERP Systems
11. Python programming

## Industrial Engineers in the Service sector

1. Healthcare: Usually industrial engineers help to optimize the healthcare industry to reduce wait times, prediction of patience incoming and outgoing, bed and floor optimization, process flow improvement, improve utilization of resources and continuous improvement.

2. Government Organizations: In big organizations such as government functions, Industrial engineers use techniques for plant or office locations, process development and optimization.

3. Banking: In Banking, mostly use of operations research methods, queueing systems, customer and service time reduction, capacity planning and staff utilization work is done.

4. Theme Park: This is big industry, here Industrial Engineering teams works on Queueing systems, wait times, Traffics and people movements and path and network optimization methods, crowd management, reduction of operational cost.

5. Logistics and Ecommerce: Typically engineers here are working for Warehouse Layout optimization, Last mile delivery optimization, Route optimization and demand forecasting methods.

# 3. Production Safety

Production floor safety is everyone's responsibility, from security guards, janitorial staff, senior leaders and industrial engineering teams. In every industrial setting production, safety is very important. It refers to the measures practices and system implementation of well-being of workers, employees, integrity of equipment, protection of environment during manufacturing processes. Industrial engineers play a vital role in production safety in analyzing and improving processes to minimize risk, reducing accidents create a safer work environment. These engineers are also responsible for having a safer factory design, designing the walkways in the factory floor, equipment, and ergonomic setting of an operator workstation, safe standard work and the safety of the ultimate customer who eventually buys the finished product.

### (OSHA) Occupational Safety and Health Administration

OSHA is a federal agency under the US Department of Labor responsible for enforcing safety standards, providing training and compliance to protect workers from occupational hazard. It began in operations because of Occupational Safety and Health Act that had been signed into law by President Richard Nixon their primary mission is to assure safe and healthy working condition for worker by setting and forcing standards, outreach education and training and assistant in safety. OSHA encourages employees to reduce or eliminate safety hazards by removing causes of hazards. Some examples are improving ventilation, having safety harnesses when you are at a certain height, how to dispose of hazardous material safely and first aid training, Gloves and use of (PPE) Personal protective equipment.

**OSHA inspection & Audit at Manufacturing Production Floor**

**Safety Manuals and Training**

Every organization should have a safety manual and should be its priority to improve safety and include bringing continuous improvement in safety. They should always encourage employees to find better ways of doing the job and the Industrial engineering outlook should start with safety. Often at production operator a few items to look for as an industrial engineer when you are conducting standards studies involving safety.

Safety Training with Production employees.

1. Safety Training
2. Lifting and work ergonomics
3. First Aid Location & training
4. Fatigue
5. PPE
6. Apparel
7. Hair
8. Tools & Equipment's safety
9. Machine activation and Deactivation
10. Hazardous chemicals
11. Safety Guards
12. Aisles and walkways
13. Machine Pinpoints
14. Ventilation

# 4. Plant Productivity & Production Cost

Productivity is the best way to measure the output created by a system with defined input and process within the system. Let's understand the production system first.

A Production system is framework used to transform set of inputs using man, raw material, labor, energy into outputs efficiently. In simple works set of factory operations which convert raw materials into finished good and products for customer.

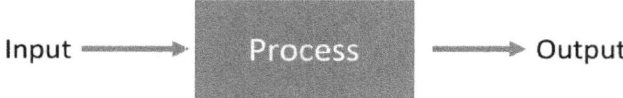

This is the smallest flow chart which can define any manufacturing plant. This exercise to convert set of inputs (raw material) to Output (Finish Products) requires resources and energy. Set of inputs

- Man
- Machine
- Material
- Money
- Management
- Information
- Energy

**Productivity** refers to the efficiency of this production system. Arithmetic equation for productivity is simply ratio of outputs to inputs.

$$Productivity = \frac{Output}{Input}$$

It's crucial for every manufacturing industry to understand how they're performing over time, year over year improvements, are the improvements implemented really making change and in production cost & generating saving initiatives. This is where productivity comes in picture for every product and production operations. There should be an uptick in productivity, which tells you that now the cost of production is going down compared to the previous trend. This tells you that productivity is improving in current improved state.

The term productivity has become wider, originally it just used to be the rate at which the worker performs his skills. The person who produces faster used to be a higher productive compared to the average. Today, productivity has various variants: productivity by departments, productivities for machines, equipment's, Labor, Shipping and logistics, Service delivery etc.

The underlined definition is same high productivity refers to doing the output work in the short as possible inputs, without sacrificing quality, and minimum wage of resources.

Some common other productivity terms which are driven from productivity are financial ratios, budgetary various, labor variance, Labor utilization, efficiency, cost of non-productivity, Capacity utilization, Manpower, production per man hour.

Productivity is the measure of how well the resources are brought together in an organization and utilized for accomplishing a set of objectives.

## Labor Productivity Example:

Consider a Car factory at three locations makes the same product A in Orlando, San Jose and New York. Standard time based on time studies it takes 1000 Man Hours to build this product.

Orlando Factory has 10 employees, and it takes 15 hours each

San Jose Factory has 5 employees, takes 30 hours each

New York Factory has 18 employees, 8 Hours each

| | *Output* | | | *Input* | *Ratio* |
|---|---|---|---|---|---|
| Factory | Standard Time (Man Hours) | # Employees | Hours Used | Actual Time (Man Hours) | Productivity |
| Orlando | | 10 | 150 | 1500 | 67% |
| San Jose | 1000 | 5 | 250 | 1250 | 80% |
| New York | | 18 | 55 | 990 | 101% |

## Productivity Benefits

There are always significant benefits from productivity improvement, starting from reduction of cost to produce the goods and the passing the low cost of product to the end consumers, also increasing the profits resulting in overall organization growth, more taxes for government & economic growth of factory location.

Between production operators and factory floor employee workforce there is always a misunderstanding about productivity, most misunderstanding is about higher productivity means higher workload, higher efforts and more profits for owners, unemployment threat or threat to their job security. These are not the correct interpretation. Productivity usually integrates between owner and worker for long term growth and creates more jobs. Higher productivity contributes to improving utilization of resources and input rather than making workers work hard. The most important part is that without changing the quality of the product.

Productivity strives to minimize human hazards. Human efforts, improve use more manufacturing equipment, effectively reduce waste from the Processes.

Productive improvement results in lower cost per unit effective utilization of resources and reduction of waste, increases factory profits and that money can be always reinvested in new technology, new equipment and new machines which further increases productivity. If a factory floor operator or Industrial engineer or Production operations Manager does not change or try new things in the production area, there is not going to be a change in productivity. You always must try new things continuously improve from current state map to future state map with planned improvements. Tweak things around and see how productivity changes and reacts, this is how productivity changes.

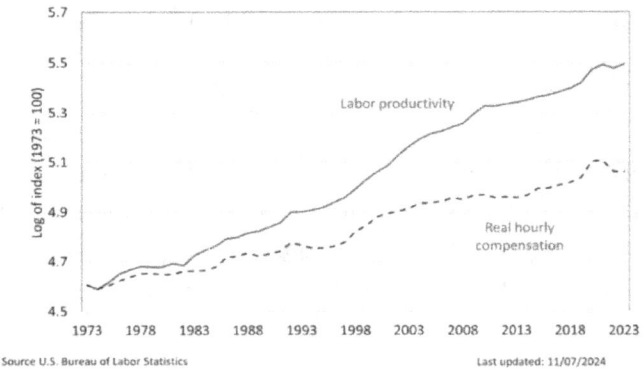

Labor Productivity and Real Hourly Compensation, Nonfarm Business Sector, 1973-2023

Improvement in productivity usually results in more output, resulting in a lowering of product cost, resulting in lowering of prices, which in turns into an increase of product demands for the good services and creates a greater employment opportunity for operators and helps increase wages and thus more bonuses for the employment and workforce. Productivity increase sets a chain reaction for better good for economy production floor operators creates more jobs

**How to increase Manufacturing Productivity**

Well, there are various ways you can improve productivity. The most important thing is continuous improvement, improving your production process step-by-step to get greater results.

Develop a robust training module for employees and create good standard operating procedures. Hire & retain the best workforce talent available and try to retain and reduce turnover. Have tools and techniques with computer systems, Internet of things (IOT) and ERP systems which will reduce systems and improve efficient operations.

This book talks about all the different ways which you can improve FACTORY EFFICENCY in future lessons and chapters. We will learn more about lean manufacturing, tools of industrial engineering, and artificial intelligence in Manufacturing. All these methods and systems are designed for improving productivity of the factory or system.

Here are a few short ways of helping and improving productivity on the production operations floor

1. Streamline Manufacturing Process and Assembly Systems.
2. Lean Manufacturing reduces waste and delays, and rework
3. Implement Standard Operating Procedures (SOPs)
4. Value stream Mapping current and Future state
5. Implement Automation as needed
6. Use IoT and ERP Systems
7. Training and Workforce skill development
8. Capacity and Demand planning
9. Incentive and recognition of the workforce
10. Just in time (JIT) Manufacturing
11. Encourage continuous improvement
12. Feedback mechanism between departments in organization and employees

**Product Cost & Pricing**

Working as an Industrial Engineer, I think it's important to understand product pricing and what strategies go behind it. This is because this will help with the importance of efficiency and productivity, and thus a role for Industrial Engineer.

**Fixed Cost**: Fixed cost are expenses that do not vary with production volume. Such as rent, lease payments, salaries and wages of full-time employment, property taxes. Usually, things you have limited control over as industrial engineer.

**Variable Cost:** Variable cost are expenses that change directly with production volume. The more you produce the higher the cost. This includes Raw materials, direct labor, packaging and shipping. These are the things which Industrial Engineer can find innovative ways to reduce.

Reduce Material cost or Optimize production process to reduce fixed labor cost.

Product costing is more complex than this but wanted to explain how productivity influences the bigger business strategies.

Fixed Cost + Variable Cost + (Profit Margins) = Product Cost.

Improvement in Productivity = Reduce Variable cost, thus increase in Profit Margins.

Note: Profits can be negative if these costs are way higher than planned product pricing.

There are more advanced demand & supply curve and microeconomics of product which dictates the market pricing of product, but As Industrial Engineer this is good enough to understand the basics of product pricing.

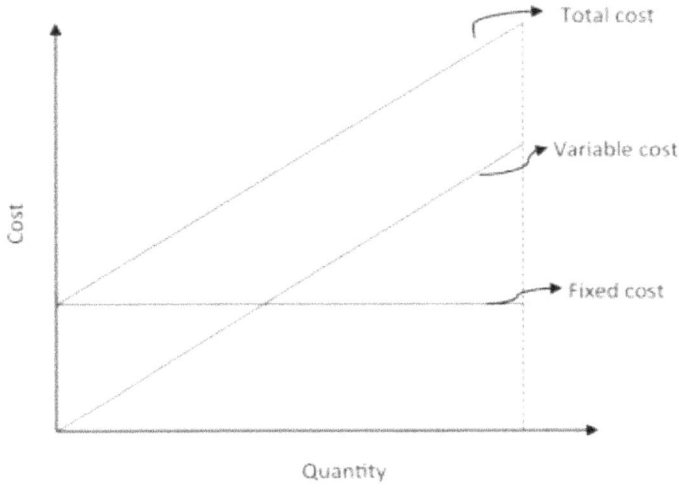

Fixed and Variable cost with Quantity

# 5. Factory Capacity Planning

Factory capacity planning is the process of determining how much production capability a manufacturing facility has and aligning it with demand forecasts. It helps ensure the factory can meet customer demands without underutilizing or overburdening resources, enabling efficient operations and cost effectiveness.

In every production system design involves planning for all inputs converted into finished goods outputs & effective management of capacity, which is the most important responsibility of Industrial Engineering and Production Management. The main objective is to match production capacity to demand.

Capacity planning is to be carried out considering future growth, expansion of market trend, sales forecast etc. This becomes simple when demand is stable. A lot of time Assembly lines are set up built and added new or merged with old, built new plant or subcontract to different vendors based on capacity planning. Lot of time Supply Chain and procurement engineering teams will also check with Tier II and Tier III companies' capacity and planning before sending them large business orders.

Capacity decisions are usually strategic decisions, usually facility capacity is expressed on volume of output per period.

Managers are concerned about capacity as it adds

- If current capacity meets market demand?
- Capacity effects cost of product
- Capacity effects scheduling system changes
- More capacity adds more investment
- Market demand and trend change

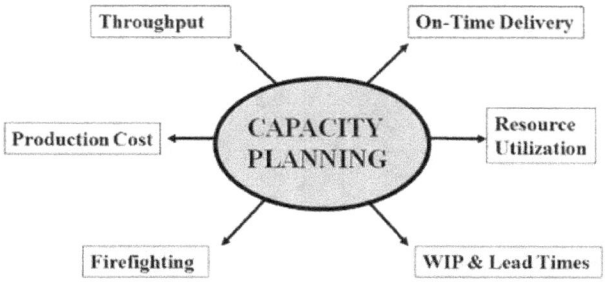

Capacity planning dependencies

Capacity planning is concerned with long term and short-term capacity needed for organization.

The following flow charts help with the best steps, usually for good capacity planning.

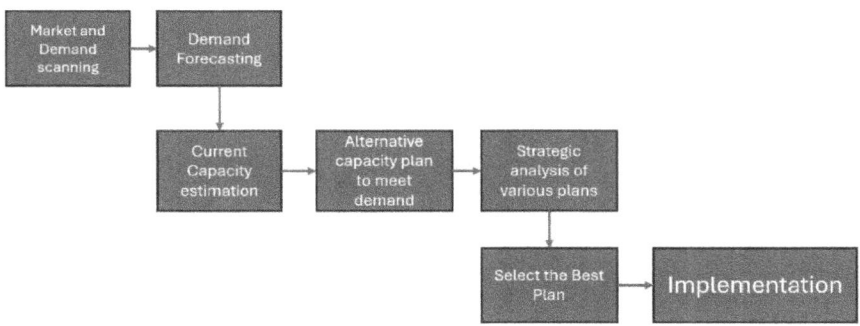

Capacity Planning Fow Chart

**Multiple Products:** Most factors will try to diversify their portfolio of products using the same factory to increase profits. The manufacturing of multiple products helps in reducing the risk of failures. Different products are at different stages of life cycle, which helps in scheduling production line to maximize capacity utilization.

**Phasing in Capacity**: Semiconductors and tech companies have high innovation every year launching new modes and products to beat market and competition, such factories and facilities need to have modular based capacity planning for 2-3 years of product life cycle or even shorter.

**Phasing out capacity**: Old factories with old machines or demand decline are usually phasing out or closing factories. Usually, these businesses and factories are reinvested to build different products or shut down in moving employees to different locations, compensating employes without hurting communities.

**Short-term Capacity strategies**: There are a lot of industries which are in cyclic period, one best example when there is urgent short term demand increase was due in COVID pandemic. Different healthcare facilities had to come up with short term strategies and innovations to modify existing manufacturing system to produce Gloves, face mask, hand sanitizer, oxygen cylinders need etc.

**Sales, Inventory and Operations Planning (SIOP)** Most factories and corporation businesses had SIOP planning meetings which has leads from Sales, Inventory, supply chain and Operations to align with supply and demand of product.

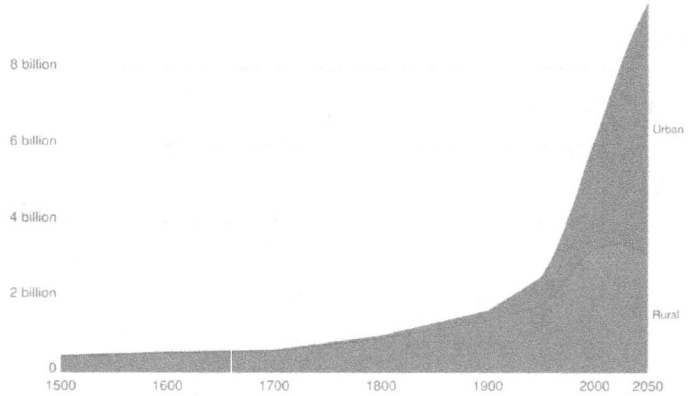

Forecast the Urban and Rural population of the world.

**Forecast of demands** heavily influence plant capacity planning. It's very difficult to forecast demand, hope with use of Artificial Intelligence there will be innovation to accurately predict product demand helping manufacturing facilities. Forecast for demand heavily depends on product season, market, pricing, product stage in life cycle, number of products.

**Plant and Labor efficiency** is very realistic to be considered while capacity planning. Standards derived by industrial engineers are based on normal working conditions, which are always hard to meet. There is always manpower with different skill levels, training, rework and machine failures in production facilities causes efficiency to drop below 100%. It's very important to use realistic values of efficiencies while considering capacity planning.

**Sub-contracting** is usually offloading workload in stage of peak demands to vendors. Make or buy analysis is done before going to sub-contracting products. Sometime part of the products is subcontracted & later assembled again in factory. This does increase the cost of product and material handling as these parts must be shipped from different plant locations and then stored in warehouse to bring it to assembly line. There are other reasons when subcontracting plays a very vital role, in my previous experience at peak demand times factory was not able to hire fast enough as labor pool in community was full and thus company must make strategic decision to subcontract and ship parts from other locations and approved vendors who have skills and capacity.

**Overcapacity** is preferred when the fixed cost is very low & sub-contracting is not possible. Time and cost required for increasing is very high. Companies can't effort to miss deliveries.

**Undercapacity** is preferred when the time to build is very short, shortage of products doesn't affect the company, technology changes very fast.

**Aggregate Planning** is an intermediate term planning, usually 2-3months for quantity and timing of output. Its process of developing analyzing and maintaining schedule that balances production capacity, inventory, workforce and demand over medium term. It also considers resources to meet production goals while minimizing cost.

**Master Production Schedule (MPS)** follows aggregate planning methods, MPS usually shows overall plan in terms of specific end items or models that can be assigned priorities. It is useful to plan for the material and capacity requirements. MPS translates aggregate planning into specific end items. MPS also prepares alternate schedules, many are computer generated using MS Excel or Gantt charts. It forms a base for inputs of material required planning. (MRP)

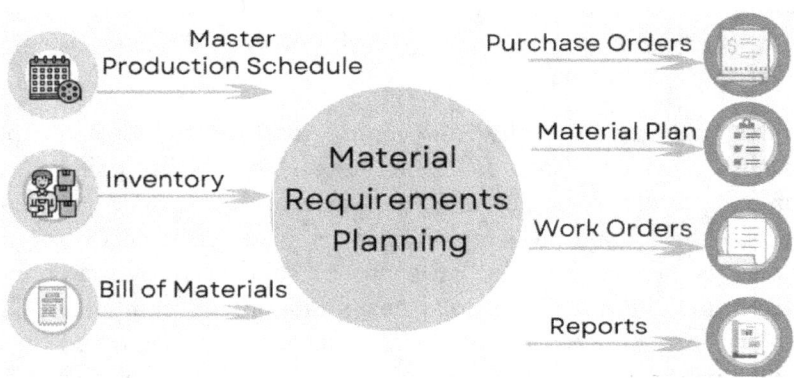

**Material Required Planning (MRP)** is an approach for determining the quantity and timing of acquisition of dependent demand items needed to satisfy master production schedule requirements. MRP Planning is a powerful tool which applied properly, helps in achieving effective manufacturing control. It's a background calculation of various smaller items based on Master production schedule, which are required to be assembled just in time at assembly stations and having inventory of it to deliver final product in time to customers.

**Bill of Materials (BOM)** identifies list of all components which are needed to assemble final product in Assembly. In Factory floor assembly line are usually multiple stations, each station has kits which have several parts, every part has unique identifying number in system which is called Bill of Materials.

# 6. Work and Method Studies

The objective of work and method studies is one the production line is set up for normal environment to determine, measure and create an engineered time standard for the job. This time standard can then be used to determine the cost of a product in terms of labor cost, material cost & fixed overhead cost of facility.

There is an option to use historical data to come up with a estimates labor cost, based on a set period such as quarter or a half a year, the cost of labor is calculated by overall output of the work divided by the actual labor hours spent, this also gives an average labor cost per unit of production. This data can be also used to determine the total cost of the finished product. Usually, industrial engineers and management are very comfortable with this number as this gives a more accurate number that reflects the actual labor cost over the past but note this also includes if there are any production downtime & non-productivity factors affecting the production floor. The drawback of this is it's a large set of data, which does not tell you standard time for individual tasks. In large-scale factories there might be a lot of non-value added work and unproductive work happening, which does increase the cost of the product. This is why Industrial Engineers and Manufacturing Engineers need to create a standard operating procedure by using time study and method studies.

Since there are a lot of employees, a huge pool of employee sometimes task is individual or a group team work. There might be a different set of skills, speed training, and the daily activities of an employee might have opportunities for wasting time doing other activities which are not part of the primary job. Sometimes if the employees finish their job earlier, the line supervisor might assign them some other work to keep them busy - this also accounts for going away from the standard work when production is slow or fast, the amount of work on everyday activity differs which creates an illusion that more time is required for the job than what is necessary.

It is important to balance the assembly lines, using detailed standard time data of every assembly line station performed by every employer, if the assembly line is not balanced without a standard time that will create an unnecessary bottleneck, thus decreasing the productivity and even the delivery of the products to end customers. This might also generate Quality defects and added rework time for assembly operator when things are not well defined.

An unbalanced assembly line can create somebody having a more workload while some other operators in the team might have less workload than needed having access free time. This can create serious bottlenecks in operations.

There are various examples in my experience where we did line optimization projects as line balancing projects create Yamami chart. We will discuss this more in the next chapters, which reduce the number of headcounts in assembly line, which were not well balanced. The standard time was not correct and needs to be fixed thus increasing productivity with the hockey stick chart eventually reducing almost 10% to 20% of headcount in the assembly line

**Time study**

Time study is a work measurement technique used in factories to determine the time required to complete specific tasks. This is well thought in Industrial Engineering classrooms in universities and schools. Time studies can be performed anywhere in restaurants, theme parks, banking and mostly for subject of this book in manufacturing of products. The purpose of time study is to define standard steps, sequence of individual sub elements of task and standard time for it. It's conducted by time study analysts or Industrial Engineering teams. In my experience every new product launched should go through time study and crease stands for every work element in it.

*Time study Observation sheet*

Camera Based time study are my favorite. I have done and conducted numerous study and stopwatch time studies. The reason I recommend camera-based time study is because you can show down speed and observe individual operators' movements, pause and replay as needed. Today's construction timelapse camera helps to add time and date on video, this also helps in Manufacturing products where a group of teams are working together to achieve output. Mostly in Fiberglass (FRP) products such as Boats, Pools and Aerospace industry.

Brinno TLC 300 Time Lapse camera for Camera based video time study

Once data is captured use excel and pivot charts to tabulate, compare and drive correlation between the time of products.

**Performance factor** is a rating factor used while conducting time studies. The subject of study is different workers which work at different speeds. The rating factor will help to raise someone at a higher-than-normal speed to 110% and slower worker at 90% which will help to come up with standard time of elemental task. It's good practice to do multiple time studies if process is a quick process.

**Personal allowance factors** always needed to be added to time studies for breaks needed for employees while completing their task. There are always different times of allowances, Break times, waiting for materials, talking to coworkers unrelated to work, on job training, safety meetings, team meetings etc. Personal fatigue and delays (PF&D) allowances may differ from different jobs.

**Value-Added and non-value-added Work:**

Based on service to customers, any work is defined into two major categories, Value added (VA) and non-value added (NVA) work. Work which is directly contributing to customer requirement by transforming raw materials into finished good are Value Added work. Certain Task do not add value directly to customers which are costly, time consuming is Non-Value-Added work.

Examples of Value-added work are Painting or finishing appearance of product, assembling parts together for final product, whereas Material movement, reworks and corrections are part of non-value-added work. There is some Necessary Non-Value-added work which does not add value to customers but added value to business which includes regular maintenance, regulatory inspections, employee training, safety and team meetings.

Basic Time Study Equipment:
- a. Stopwatch
- b. Time study form or software
- c. Video Camera if you plan to record it.
- d. Clipboard and pencil
- e. Tape Measure to provide dimension of work or sketch of work location

Time Study Sketch on Factory Floor

**Lead time** is the total time it takes for a process or activity to be completed from the time it received ordered to the final delivery of product. This is important for designing assembly line systems around lead time. It's a critical metric in manufacturing, supply chain management and operations, as it affects customer satisfaction, production planning & inventory.

**Cycle time** is the total time required to complete one unit of production. Cycle time includes setup time, processing time, waiting time or moving time.

$$\text{Cycle Time} = \frac{\text{Net Production Time}}{\text{Number of Units Produced}}$$

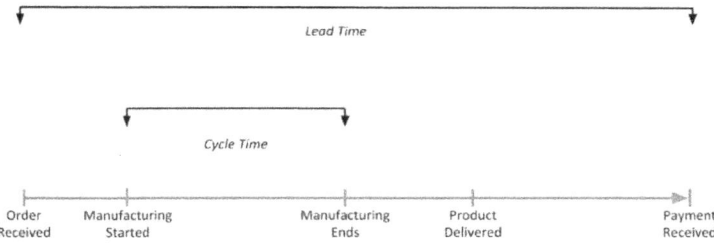

**Takt Time** is rate at which products must be completed to meet customer demand with specific timeframe. In Assembly lines, the line moves with takt time. It can be in minutes, seconds or even in days based on demand requirements.

$$\text{Takt Time} = \frac{\text{Workable Production Hours}}{\text{Units Required (Customer Demand)}}$$

**Standard Operating procedures (SOPs)** in Manufacturing are set of steps-by-steps instructions that outline to complete the routine task. They are detailed, written documents that are required to perform a specific process for consistency and efficiency. SOPs are used across industries to ensure operations are carried out uniformly, safely, and in compliance with organizational or regulatory requirements. SOPs are used for new employee training also help in keeping consistency of product quality thus increasing efficiency of factory. It's important to maintain editing and modify SOPs with time in every industry as the factory continuously improves.

# 7. Tools in Industrial Engineering

The idea behind this chapter is to discuss various tools and techniques that Industrial Engineering team will use in Manufacturing and production environment to increase plant productivity. In previous chapters we discussed a few items such as time studies and Method studies, coming up with standard work elements and standard operating procedure.

Method Study: To Establish a standard method to perform a job or operation and establish a layout of production facility.

Time Study is a work measurement technique we covered in previous chapter to establish standard time for a job

Motion Economy: To analyze the motions employed by operators to do the work, further this can be classified into value added and non-Value-added work

Financial and Non-Financial Incentives: Helps to evolve at a rational compensation for the efforts of workers. Incentive programs for workers

Value Analysis: It ensures that no unnecessary cost is built into the product and helps to enhance the worth of the product

Production, Planning & Control: Includes the planning for resources, proper scheduling and controlling production activities, the right quantity, quality of product at predetermined time and pre-established cost.

Inventory Control: To find the economic lot size and the reorder levels for the item to be available to production at the right time and quantity.

Job Evaluation: A technique used to determine relative worth of a job to aid in matching job and personnel at sound wage policy.

Material Handling Analysis: To scientifically analyze the movement of materials to eliminate unnecessary movement and enhance the efficiency of material handling.

Human Engineering & Ergonomics: Ergonomics is concerned with studying of relationship between worker and its working conditions to minimize mental and physical stress.

Operations Research: Aid to arrive at the optimal solution to the problems based on the set objective and constraints imposed on the problem.

System Analysis: Is a study of various sub systems and elements that make a system achieve greater efficiency and effectiveness.

**Process Flow charts**: This is a very helpful tool to understand large and complex processes and systems in how they function. Manufacturing process flow design is a method used to evaluate specific processes that convert raw material into multiple stages of actions into finished goods.

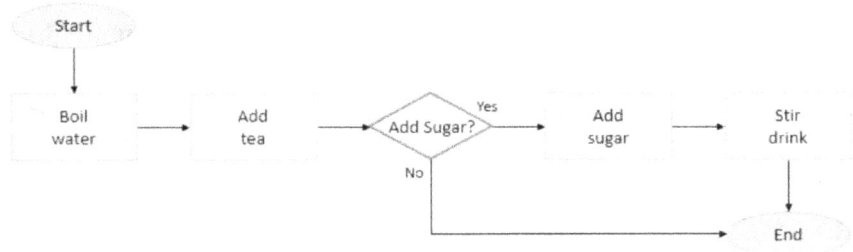

Example of process flow chart

**SIPOC diagrams:** A SIPOC diagram is a high-level process mapping tool used in Six Sigma and process improvement methodologies. It's an add on to process flow map, but in structured way to analyze process by key elements: Supplier, Inputs, Process, Outputs and customer

S – Suppliers   Entities (internal or external) providing inputs to the process.

I – Inputs   Resources, materials, or information required to execute the process.

P – Process   The high-level steps or activities that transform inputs into outputs.

O – Outputs   Products, services, or results delivered by the process.

C – Customers Recipients of the process outputs, either internal (departments) or external

Simulation & CAD: Simulation involves using digital models to replicate, actual to analyze and optimize real world manufacturing problems. CAD software's should be used for factory flow and layout planning respectively.

**SIX SIGMA** is a data driven methodology which focuses on process quality by identifying and eliminating defects. Defects cause rework and loss to product and organization. Six Sigma methods use statistical tools and techniques to achieve near-perfect results, targeting a defect rate of 3.4 defects per minion opportunities.

Six sigma is widely applied to optimize process, reduce waste and enhance product quality. These principles are customer focus, data-driven decision making, process optimization, collaboration and continuous improvement culture. American society of Quality (ASQ) offers certification for professionals White Belt, Yellow Belt, Green Belt, Black Belt & Master Black Belt.

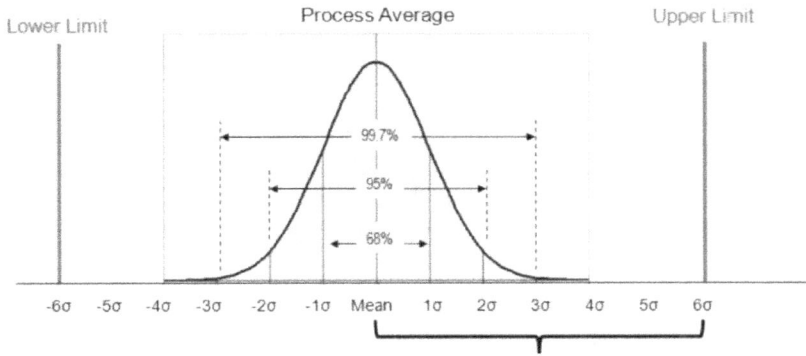

Six Sigma Normal Distribution

**DPMO:** Defects Per Million Opportunities is a metric used in Six Sigma to measure quality performance of a process. Its number of defects in process per one million opportunities.

**Formula for DPMO**

$$\text{DPMO} = \left( \frac{\text{Number of Defects}}{\text{Total Units Produced} \times \text{Number of Opportunities per Unit}} \right) \times 1,000,000$$

DIMAC is a structured, data driven problem solving approach using six sigma to improve existing processes by eliminating defects and reducing variability. The acronym DMAIC stands for Define, Measure, Analise, Improve and Control.

Tools used in DIMAC

| | |
|---|---|
| Define : | SIPOC, Voice of the Customer (VOC), CTQ Tree |
| Measure: | Process Mapping, Gage R&R, Data Collection Plan |
| Analyze: | Fishbone Diagram, Pareto Chart, Hypothesis Testing |
| Improve: | Design of Experiments (DOE), FMEA, Poka-Yoke |
| Control: | Control Charts, Standard Operating Procedures, Audits |

## 7 QUALITY TOOLS

Seven Quality tools are simple and effective ways used for problem solving and process improvement in quality management.


1. Cause-and-Effect Diagram : Identify and organize potential causes of a problem.

2. Check Sheet: Collect and organize data systematically.

3. Control Chart: Monitor process variation over time.

4. Histogram: Visualize frequency distribution of data.

5. Pareto Chart: Prioritize problems based on their significance.

6. Scatter Diagram: Analyze relationships between two variables.

7. Flowchart: Map and understand process steps.

# 8. Lean Manufacturing

Lean Manufacturing term as it states, making production lean by removing waste. This term was coined in book The Machine that Changes the World, in which it compared Japanese and American companies. The Toyota Production System (TPS) was the first system working according to Lean Manufacturing. To Implement Lean and continuous improvement culture, Leadership and Top Management should prioritize it, employee empowerment, visible matrix and regular lean manufacturing training withing operators. Encourage small improvements across all levels of the organization through Kaizen (continuous improvement) while consistently celebrating success and adapting to changing needs. There needs to be commitment from Leadership and executive GOALS for lean Manufacturing development and engagement of employees.

Notable benefits of Lean Manufacturing are reducing waste and rework, Improved efficiency, cost savings, Higher quality, increase flexibility and employee engagement. Lean Manufacturing principles have now moved into healthcare, retail, ecommerce and software development IT industries.

Lean focuses on waste reduction, while six sigma is based on defect reduction. Together they are often referred to as Lean Six Sigma methodologies. In this chapter we will explain a few basic concepts and tools used in Lean Manufacturing.

**Value Stream Mapping:** Is a lean tool used to visualize, analyze and improve flow of materials and information in systems. It's more detailed than a process flow map. VSM has information about number of operators, takt time, cycle time and lead times in building value stream maps.

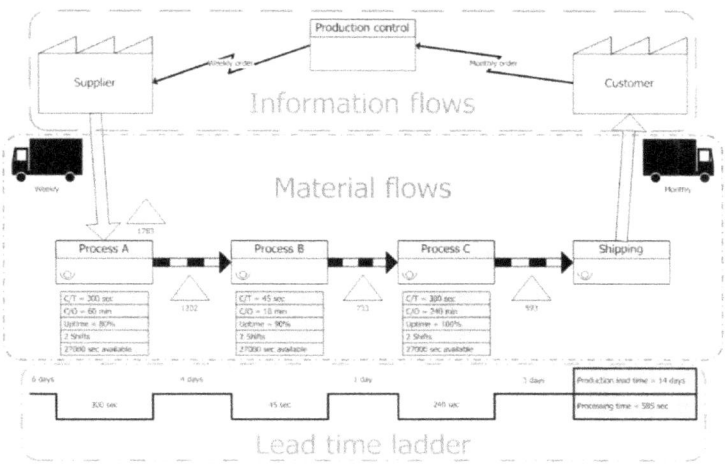

**KAIZEN** refers to ongoing or continuous improvement. Kaizen can be broken down into two Japanese words: Kai, meaning Change Zen meaning good. It translates into constant improvement for good. It's a concept of doing better every day, with everyone and everywhere. Kaizen culture always thrives to try new things and challenges the phrase, that's how we did it always. It tried to eliminate silos, egos, and waste through micro changes and instead aspires to efficient and standardize operations. Kaizen is a method of continuous improvement founded on premise that tiny, continual positive adjustment may yield substantial results. Often it is established on collaboration and commitment rather than techniques that employ drastic or top-down reforms to achieve transformation. Kaizen has the potential to improve every department from sales, finance to customer service.

Steps in Kaizen event:

- Define scope and problem statement
- Gather information
- Draw current state map
- Identify waste
- Design future state map
- Implement Change
- Review and Refine

DOWNTIME is an acronym representing 8 different types of waste. While working on any tools above, always have lookout for waste and plan to reduce or remove this waste completely.

## The "8 Wastes"!

**Defects**
Errors
PA forms, AdComp forms, Incomplete information for Grants submission

Overproduction
Doing more than needed
Extra reports, Unnecessary info. sent automatically, printing in advance

Waiting
Waiting or Delays
Waiting for information, report, answer, approvals, signatures, etc.

**Not Utilizing Employees**
Ideas and skills not used
Not recognizing employees as best source for fixing issues

Transport
Movement of people or material
Transport between campuses, Movement of files to different locations

Inventory
Too much material
Buying in bulk, more servers than required, supplies, equipment

Motion
Movement by workers
Searching for supplies, items needed not close by, always looking in shared drives

Extra Processing
Re-dos
Unnecessary approvals, rework, same data required in multiple places or systems

5S is a workplace organization methodology that enhances efficiency, safety and productivity by maintaining order and cleanliness. The 5S system is derived from 5 Japanese words that each represents a key concept for workplace organization.

| Japanese Term | English Translation | Description |
|---|---|---|
| Seiri | Sort | Identify and remove unnecessary items from the workplace. |
| Seiton | Set in Order | Organize tools and materials for easy access and retrieval. |
| Seiso | Shine | Clean the workplace to maintain a safe and efficient environment. |
| Seiketsu | Standardize | Establish standards to sustain the first three steps. |
| Shitsuke | Sustain | Develop habits to ensure continuous adherence to 5S. |

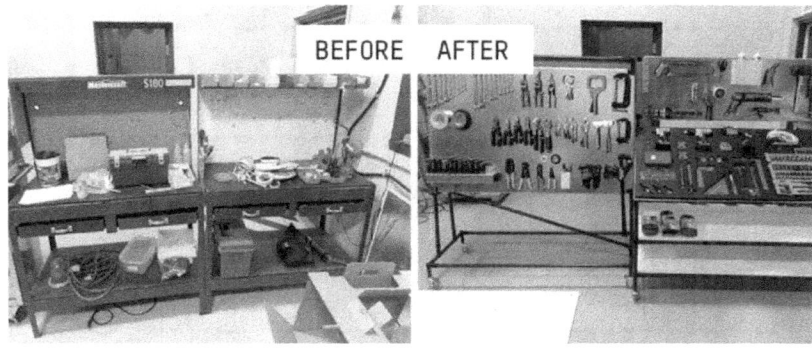

## Just in Time (JIT)

Just in time is a system of production that makes kits and parts delivered to the assembly floor when it's needed right on time. JIT and Jidoka are the two pillars of Toyota production system. JIT relies on heijunka as foundation and it is comprised of three operating elements: the pull system, takt time, and continuous flow. The idea of JIT is credited to Kiichiro Toyoda in mid 1930s.

**Toyota Production system (TPS)**

The Toyota Production system (TPS) is a manufacturing philosophy developed by Toyota that focuses on efficiency, quality and waste elimination. It's built around two core principles of Just in time and Jidoka (Automation) with human oversight to detect and address defects immediately.

TPS Promotes pull production system, which we will discuss in upcoming chapters of assembly line systems.

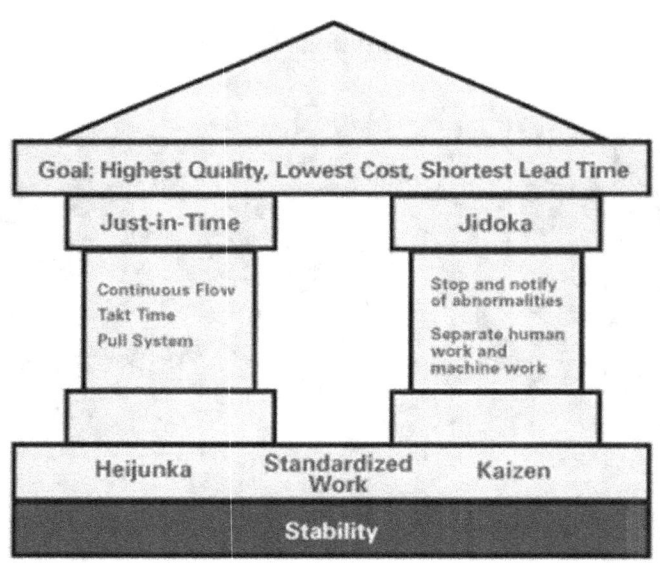

Toyota Production System "House."

House of Toyota Production System (TPS)

Root cause analysis is a systematic process used to identify the primary cause of a defect or problem and issue. RCA creates a step of actions enabling effective corrective actions to prevent recurrence. It's a systematic process to identify and understand underlying causes of problems or defects in manufacturing. RCA usually helps with reducing rework, improving cost and productivity.

Some Tools used for Root cause analysis are.

**5 WHY** is a practice of asking why repeatedly whenever a problem is encountered to get beyond the obvious symptoms to discover the root cause. It's important to implement countermeasures to keep the problem from occurring again in the future.

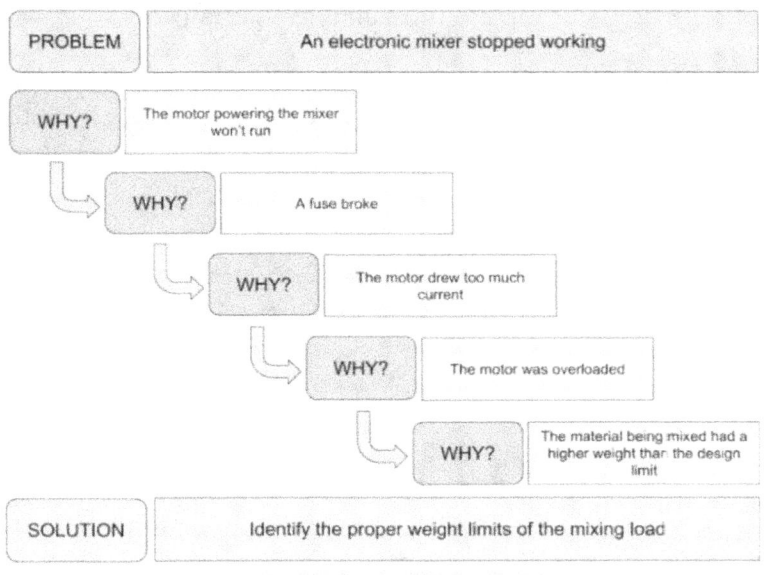

**Root Cause Analysis**

Example of 5 Why

**Ishikawa Diagram** also known as Fishbone diagram, is a visual tool that helps identify and analyze the root cause of a problem. It's a structured approach that can help teams diagnose problems and reach consensus on their causes. It tried to divide causes of problem into Man, Material, Machine, People, Environment, and method practices.

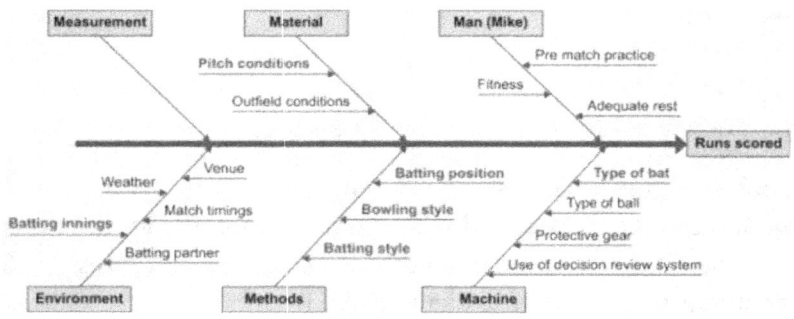

Example of Fishbone Diagram

**Pareto Chart** is a bar graph charting method using data collection and showing relative importance of problems. Length of the bar represents frequency and cost and are arranged with longest bars on left and shortest on the right.

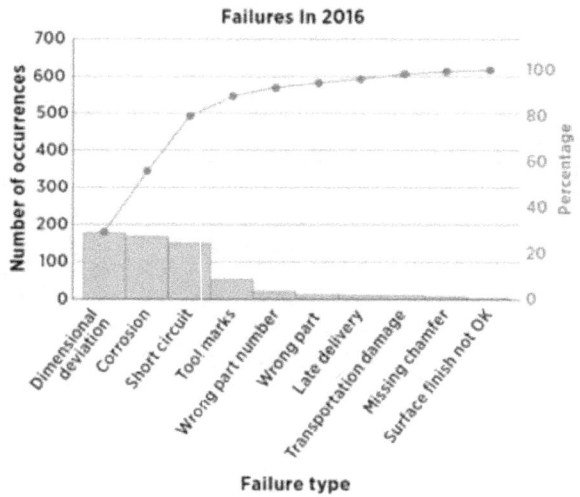

**Andon System** is an alert system on the manufacturing production floor that notifies operators that product issues have been detected. Systems can record and tract various downtime caused in assembly line, team can then use tools like pareto chart to understand issues which are repeating and ultimately eradicate them using RCA principles.

**Example of Andon lights**

**Gemba walk** is a management strategy when teams and cross functional teams meet at location where the work is done to understand process and identify opportunities of improvement.

# 9. ERP Systems

**Enterprise resource planning (ERP)** is a software application that organizations use to manage day-to-day business activities such as accounting, procurement, project management, risk management and compliance and supply chain operations. In Manufacturing and factory floor all Purchase orders, options, special instructions, inventory, kits, labor production planning and control are LinkedIn and interconnected to system. The ERP system ties together a multitude of business processes and enables the flow of data between them. By collecting an organization shared transactional data from multiple sources, ERP systems eliminate data duplications.

ERP systems are critical for managing thousands of businesses of all sizes in all industries. To these companies ERP is the heart of their operations.

```
MAIN                            i5/OS Main Menu
                                                        System:   OSYS1
 Select one of the following:

     1. User tasks
     2. Office tasks
     3. General system tasks
     4. Files, libraries, and folders
     5. Programming
     6. Communications
     7. Define or change the system
     8. Problem handling
     9. Display a menu
    10. Information Assistant options
    11. iSeries Access tasks

    90. Sign off

 Selection or command
 ===> _

 F3=Exit   F4=Prompt   F9=Retrieve   F12=Cancel   F13=Information Assistant
 F23=Set initial menu
 (C) COPYRIGHT IBM CORP. 1980, 2005.
```

AS400: ERP system example

ERP systems have multiple functions and systems connected which helps for processing data smoothly. Few notables are as follows:

54

Production Planning: Creating efficient production schedule and Master Schedule dates in system which shows date and time what a product is slowing in the system of assembly steps and coming out as finished goods.

Material Management: It monitors in real time how many parts are in the warehouse, part number, bill of materials and current cost. Supplier information and lead times from supplier.

Shop Floor Control: Monitoring and tracking Assembly line product location Realtime, tracking labor used, production bottlenecks.

Inventory control: Maintain accurate inventory levels, safety stock.

Quality Control: Managing quality inspections, holds and warranty information.

Sales and Order Management: Processing customer orders, managing sales forecasts and coordinating delivery schedules.

Supply Chain Visibility: Tracking finished goods through vendors and supply chain. Cutting Purchase orders to vendors

Finance Management: Integrating production cost with accounting functions, generating financial reports and managing budgets.

Example of ERP System

The flow chart below will show a typical assembly station process and different triggers and actions that ERP system will do while Station is complete, Station is in progress and when work is in Queue for next station.

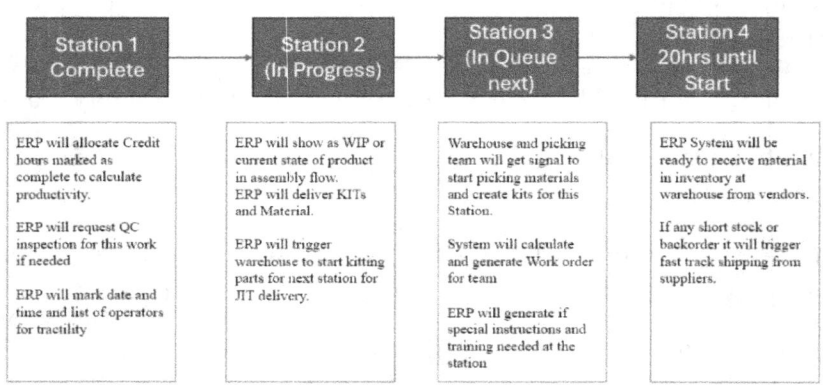

# 10. Assembly line Design

Assembly line is a series of workstations to assemble products. Designing an efficient manufacturing assembly line is an old problem yet also a new problem. In the past several enormous knowledge and design methodologies have been developed. Product innovation, manufacturing technology, robotics, product demand and product lifecycles often result in new design problems. Line design therefore plays important role in Manufacturing and factories. Assembly line design usually is derived from advanced mathematics, existing literature, design environment, new product and market demand fluctuations. Some problems while designing assembly lines are, Unknowns solved by generality. Uncertainty solved by more flexible assembly lines, Variation solved by adaptability of Assembly line, complexity solved by deep analysis and communication.

There are various line components to look for while designing Assembly line

1. Process Design/ Process Flow
2. Line Balance
3. Test strategy
4. Yield Management
5. Material Handline
6. Maintenance policy
7. Work in Progress Management
8. Parts procurement
9. Just in Time / Kitting
10. Line Size and Layout
11. Information Systems

**Littles Law in Assembly Line System:**

Littles law is part of Queuing theory, is a theorem by John Little which states that the long-term average number L of customers in a stationary system is equal to the long-term average effective arrival rate λ multiplied by the average time W that a customer spends in the system. Expressed algebraically the law is

$$L = \lambda W.$$

This can be converted into Assembly line function by factors such as Throughput, Work in progress and cycle time.

TH = WIP / CT

where TH is expressed throughput in terms of items/unit time, WIP is work-in-process expressed in terms of items and CT is cycle time expressed in terms of time units/item.

PUSH & PULL Systems

Every manufacturing facility you need to produce what your customers want. The challenge is knowing exactly what they want, when they want it. In general every factory will have mix combination of push or pull. This can also be analyzed between departments in the factory.

Push system means company produces goods according to a demand forecast. Its also known for Make to stock manufacturing. There are open goods which have slow change of demand fluctuations Ex: Food pharmaceutical, household chemicals, electronic devices, bakery.

Pull system is a lean manufacturing strategy where goods are produced according to actual demand that forecast. In this kind of manufacturing system, companies only keep as much inventory and produce as much is needed to respond to existing customer orders.

Great example is JIT( Just in time) to schedule the process of material exactly when its needed. Another approach is Kanban's, which triggers production and inventory movements when needed.

| PUSH SYSTEM | PULL SYSTEM |
|---|---|
| Production Based on Demand | Production based on Order |
| Forecasting | Actual data |
| High Inventory | Low Inventory |
| Products with high lead times | Products with less lead times |
| High risk of waste | Low waste |

**Tyles of assembly line configurations:**

There are various layouts possible while designing the assembly line, based on number of stations, cycle time and demand calculations: -

Serial Line: Single Station are arranged in a straight line along a conveying system. Each station performs one or more tests on the partially finished product and can be a simple unit of a complex system.

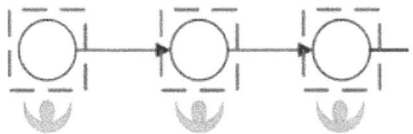

Series assembly line

**U Shaped Line** workers are placed in center of U and can monitor each other's progress and collaborate easily whenever required. Thus, workers acquire multiple skills leading to higher motivation, improved quality of products and increased flexibility.

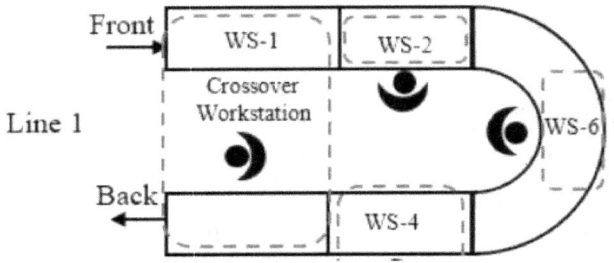

*Example of U-shaped Assembly Line*

Parallel Stations: With High production rates, the longest task time sometimes exceeds the specific cycle time. A common remedy is to create stations with parallel or serial posts, where two or more workers perform an identical set of tasks.

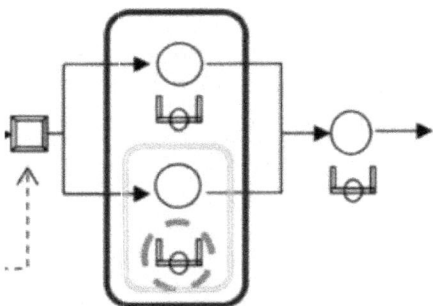

*Example of Parallel station*

Parallel Lines are duplicate of entire assembly line parallel when demand is high. The advantage is shortening the assembly lines but requires more equipment and tooling. If failure occurs at the given station another line can continue to run.

Workcenters is a designated area in the production facility where specific tasks are performed. Its critical component of assembly line needs to be designed with all tools needed, ergonomic, material handling needs.

**Line balancing** is the process of distributing task evenly across workstations in an assembly line to ensure each workstation operates at a similar cycle time. They avoid bottlenecks and idle time, improving overall efficiency. Key Objectives of line Balancing is to Minimize Bottlenecks, Reduce idle time, Meet takt Time.

Steps in Line Balancing:

1. Analyze task
2. Determine Takt time
3. Assign task to workstations
4. Adjust for constraints
5. Iterate and Optimize

**Yamazumi** is a chart, visual tool. Yamazumi starts for stck up, or Bar chart. It's a visual way to display workload of each station or operator in a stacked bar chart format. It's commonly used with line balancing.

Components of Yamazumi Chart:

1. Bar representation of Task: Each bar shows the time taken for tasks at a workstation.
2. Color Coding: Task may be color coded to include
    a. Value added Tasks
    b. Non-Value-added Task
    c. Support task

3. Comparison Against Takt time: A Horizontal line drawn to represent Takt time. Bar exceeds the line indicate imbalance or potential bottlenecks.

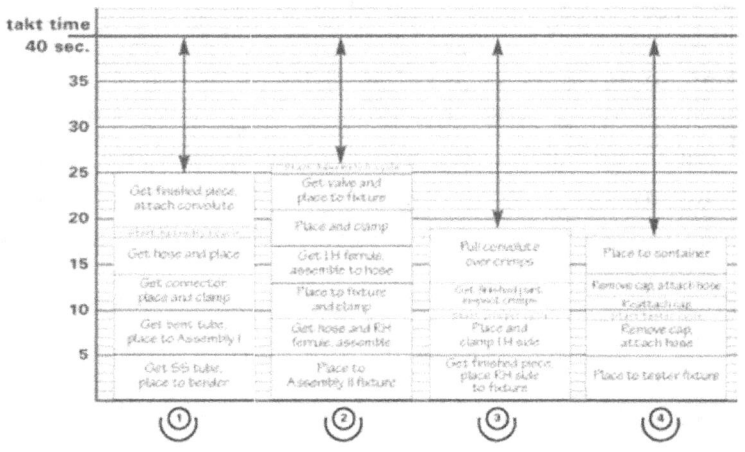

*Example of Yamazumi chart*

Statistical Function are variation in quantities in manufacturing, although in six sigma is all about reduction in variation, but in some industry which is totally dependent on craftmanship, statistical variation is highly possible. To take this into consideration, sometimes in line balancing and yamazumi it's important to have some room for fluctuations and sometimes defects.

**Heijunka** is a lean Manufacturing technique used to level production by distributing work evenly over a set period, aligning production with customer demand while minimizing waste. It is a core concept in the Toyota Production System (TPS), essential for achieving Just-In-Time manufacturing.

a. **Leveling Volume:** Adjust production to match fluctuating customer demand.
b. **Leveling Variety:** smooth production mix to reduce variability.

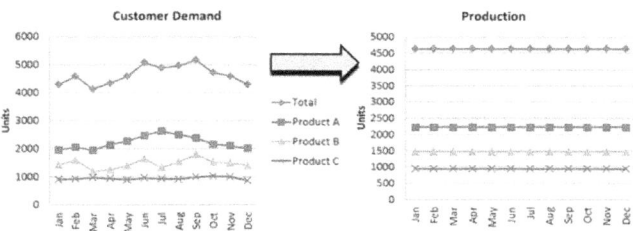

Example of Heijunka

Based on scheduling, variation , productivity and line balancing there should be a good labor planning done and balance the capacity of production.

**Batch & One-piece flow Production:**

Single production refers to one individual product at a time while batch production means producing a set of quantities in groups.

**Work in Progress** refers to partially completed goods that still are undergoing in production. These items are neither raw material nor finished products, they are typically tracked as part of the company's inventory.

**Advance algorithm in assembly** line design leverages computational technique to optimize layout, balance workloads, minimize costs, and enhance efficiency. Few methods if you want to learn this in deeper level are

1. Linear Programming
2. Dynamic Programming
3. Genetic Algorithm
4. Particle Swarm Optimization
5. Ant Colony Optimization
6. Constraint Programming

**Mix model lines- (MMALs)** Mix-Model Assembly lines where multiple products variants are produced on same line with varying volumes. These types of lines are commonly found in industries such as automotive manufacturing, electronics and consumer goods. The challenge in Mixed Model assembly line is balancing task, task variability, cycle time balance, workstation assignment, inventory management.

Steps to design MMALs

    a. Define product variant and Product Family.
    b. Task Breakdown of Product Family
    c. Value stream Map and Process Map
    d. Task Breakdown
    e. Balancing Line
    f. Sequencing
    g. Optimization

**Queuing Theory** is mathematical study of waiting in lines or Queues. It is widely used to model systems where there are limited resources that must service incoming requests of task, and it aims to predict queue lengths, waiting times, and another key metrics to optimize system performance.

    a. Arrivals
    b. Service Mechanism
    c. Queue Discipline
    d. System capacity
    e. Queue length
    f. Waiting time

## Common Queueing Models

**M/M/1:** A single-server system with exponential arrival and service rates.

**M/M/c:** A multi-server system with exponential arrival and service rates.

**M/G/1:** A single-server system with a general service time distribution.

**G/G/1:** A single-server system with general arrival and service time distributions.

**M/D/1:** A single-server system with deterministic (constant) service times.

**M/M/$\infty$:** An infinite number of servers, where each arrival gets immediate service.

These models can be used in advance practice for Manufacturing assembly line, workstation optimization methods.

# 11. Bottleneck Analysis

All plants and factories have bottlenecks, there is never a perfect and balanced plant. It's part of recognizing and analyzing them and eradicating it. A manufacturing bottleneck refers to a point in production process where the flow of material or task is limited, resulting in delays, reducing throughput and inefficiencies. Bottleneck can occur at any stage of manufacturing, weather in assembly, material handling, testing or finishing.

Causes of Bottleneck

1. Capacity constraints
2. Inefficient workflows
3. Material Handling issues
4. Quality Control and Inspection
5. Operator skills
6. Technological Limitations

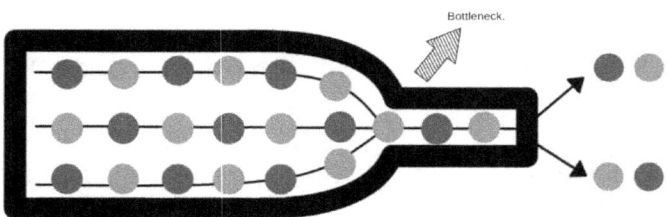

Example of bottleneck

**Recognize and Identify Bottleneck**

One of crucial steps is bottleneck analysis and accurately identifying bottlenecks or constraints within a process.

1. Monitor Throughput and cycle times
2. Utilize Queueing Theory
3. Production simulation
4. Visual Management tools

5. Root Cause Analysis
6. Visual inspection and WIP increased signals
7. Increase in inventory at station signal

Impact of Bottleneck

a. Reduce throughout
b. Increase Lead times
c. Higher Inventory Levels
d. Higher Operating Costs
e. Decreased Customer Satisfaction

Strategies for Removing Bottlenecks

1. Identify and Prioritize Bottlenecks – Use Pareto Analysis to focus on 20# of process that contribute 80% od delays.
2. Increase Capacity at the Bottlenecks- Add more stations or workers, machine.
3. Process Improvement (lean Manufacturing) VSM or Kaizen
4. Balance Workload- Line balancing or Task reallocation
5. Buffer Management- Create Buffers stock between bottlenecks. Implement Just in Time at bottleneck stations.
6. Reduce setup times – Lean principle of SMED etc
7. Cross train employees
8. Improve Maintenance and Reliability – Reduce downtime ii production
9. Use Automation


Theory of constraints (TOC) is a systematic approach to identify and address bottlenecks.

1. Identify constraints—Slowest process in the production chain/
2. Exploit the constraints—Make the most of the current capacity at bottleneck.

3. Subordinate everything else—Align other process to support capacity of the bottleneck
4. Elevate the constraints—Invest in resources or improvement
5. Repeat the process- Once a bottleneck is removed a new constraint may emerge, repeat the process.

Manufacturing plants should not focus on balancing individual workstations or resources for efficiency in isolation, but rather, they should focus on optimizing the performance of the entire production system by addressing the constraints, bottlenecks that limit throughput.

# 12. Engineering Economics

Engineering Economics is critical for decision making in engineering projects and plays central role in ensuring that resources are utilized effectively to achieve the desired outcome. It can be summarized by a few key items

    a. Informed Decision Making
    b. Efficient Resource Allocation
    c. Cost Estimation
    d. Balance Technical and Financial Goals
    e. Understanding economic gains
    f. Risk Assessment
    g. Maximizing Return on Investment
    h. Policy and Strategic Planning

Principles of Engineering economic Analysis are: -

1. Money has a time Value
2. Make investments that are economically justified
3. Choose the mutually exclusive investment alternative that maximizes economic worth
4. Two investments alternatives are equivalent if they have same economic worth
5. Marginal revenue must exceed marginal cost
6. Money should continue to invest as long as each additional increment of investment yields a return that is greater than investors' time value of money
7. Consider only differences in cash flow among investment alternatives
8. Compare investment alternatives over a common period of time
9. Risks and return trend to be positively correlated
10. Past costs are irrelevant in engineering economic analysis, unless they impact future cost.

**Time Value of Money (TVOM)** is fundamental concept in engineering economics and finance that recognizes that the principle that money available today is worth more than the same money amount in the future, due to its potential earning capacity. This concept is essential in evaluating investment opportunities, project feasibility, and long-term financial decisions.

**Net Present Value** is a core concept in finance and engineering economics that measures the profitability of a project or investment. It represents the difference between the present value of cash inflows and the present value of cash outflows over a specific period, using a discount rate to account for the time value of money (TVOM).

**Cash flow diagrams** visually represent income and expenses over some time interval. The diagram consists of a horizontal line with markers at a series of time intervals. At appropriate times, expenses and costs are shown.

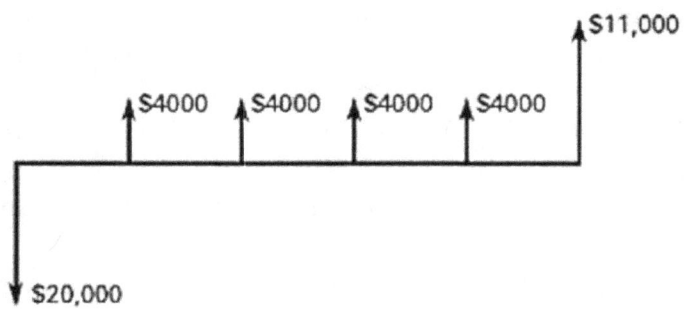

**Example of Cash Flow Diagram**

**Break Even point (BEP)** is the moment a company's operations stop being unprofitable and start to earn a profit. The breakeven point is the production level at which total revenues for a product equal total expense. The breakeven point can also be used in other ways across finance such as in trading.

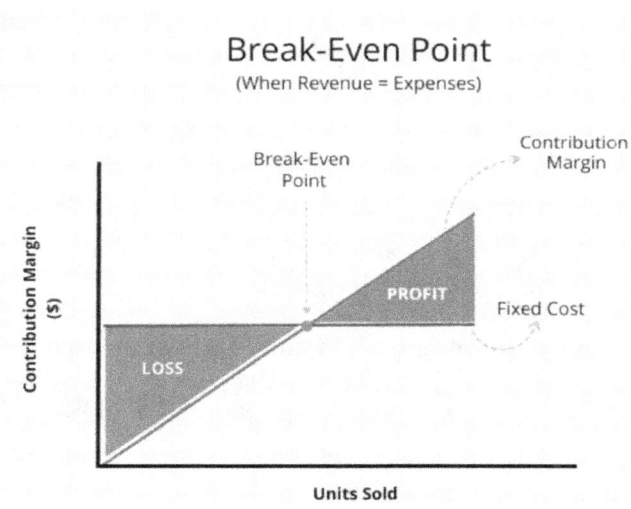

**Example of Break Even Point**

**Return of Investment (ROI)** the financial gain a manufacturing company receives compared to the cost of an investment made, typically calculated by dividing the net profit generated by a new machine, process, or technology by the initial investment cost, expressed as a percentage; essentially, it measures how much profit a company earns for every dollar invested in manufacturing operations.

**ROI = (Net benefits - Costs) / Costs x 100%**

**Profit Margins** the percentage of revenue that remains after a company has paid for the costs of producing a product. It's a key indicator of a company's profitability, and can vary widely depending on the industry, company size, and market conditions.

A good profit margin for manufacturing companies is generally between 10% and 20%. However, a 5% net profit margin is considered low, while 20% is considered high

$$Profit Margin = (Net Profit/Revenue) \times 100$$

**Eisenhower Matrix**

The Eisenhower Matrix is a tool that helps you organize and prioritize tasks by urgency and importance. It divides tasks into four quadrants, and each quadrant has a specific action to take:

- Urgent and important: Do these tasks immediately
- Urgent but not important: Delegate these tasks to someone else
- Not urgent but important: Schedule these tasks for later
- Not urgent and not important: Delete these tasks

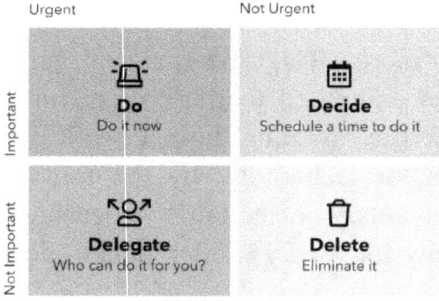

**Example of Eisenhower decision Matrix**

**New Product Development** is a process of bringing new products from an idea or concept to market. It involves several stages from identifying customer needs and generating ideas to designing, prototyping, manufacturing and launching the product.

**Industrial Engineering Role in New Product Development:**

IEs plays a vital role in new product development by optimizing processes, resources and systems.

1. Product design Optimization
2. Process Design and Optimization
3. Cost Analysis and Reduction
4. Supply Chain and Logistics Planning
5. Prototyping and Testing
6. Lean and Agile Methodologies
7. Quality Assurance and Control
8. Workforce Planning and Training
9. Project Management
10. Project Management

**Industrial Engineering Tools and Techniques in NPD**

1. Value Engineering (VE): Focuses on improving product value by optimizing functionality and reducing costs.
2. Design of Experiments (DOE): Ensures robust product design through systematic testing.
3. Simulation and Modeling: Uses tools like CAD and process simulation software for design and process evaluation.
4. Six Sigma: Enhances product and process quality by minimizing variability.
5. Time and Motion Studies: Improves efficiency in assembly and production workflows.

# 13. Finance Basics for Industrial Engineers

Finance is a crucial area of knowledge for industrial engineers, especially for those involved in process optimization, project management and strategic decision-making. Understanding basic finance concepts tailored to manufacturing setting.

The GOAL for every single factory is to make money. Industrial Engineers make find ways to increase profits by designing efficient systems or indirectly make things working towards goals.

A few important definitions for the book **'The Goal: A Process of Ongoing Improvement'** that every Industrial Engineer should think while working on any project: -

**Productivity** is the act of bringing a company closer to its goal. Every action that brings a company closer to its goal is productive. Every action that does not bring a company closer to its goal is not productive. The goal is to make money. Starting differently, the goal is to increase throughput while simultaneously reducing both inventory and operational expenses.

- Throughput is the rate at which the system generates money through *sales*.

- Inventory is all the money that the system has invested in purchasing things that it intends to sell.

- Operational expense is all the money the system spends to turn inventory into throughput.

- Fluctuations don't average out. They accumulate. It's an accumulation of slowness because dependencies limit opportunities for faster fluctuations.

More than 80%-90% of saving on any projects is going to be designing Robust line, projects such as Line Balancing, Kitting—rest 10%-20% will be coming from Lean principles such as 5S, Waste reduction etc. This is very important, now a days Teams and Manufacturing focus on LEAN first and fails to design a robust system of assembly line, thus it Projects do not impact savings or reflect in companies P&L Statement.

A **Profit and Loss (P&L) Statement**, also known as an **Income Statement**, is a financial document that summarizes the revenues, costs, and expenses incurred during a specific period in a manufacturing business. It shows whether the company is making a profit or incurring a loss, making it an essential tool for evaluating financial performance.

1. **Revenue (Sales)** Total income generated from selling manufactured goods.
   includes revenue from primary operations (product sales) and any other income sources (e.g., scrap sales).

2. **Cost of Goods Sold (COGS)** The direct costs of manufacturing products, including:
   a. **Raw Materials:** Cost of raw materials used.
   b. **Direct Labor:** Wages for workers directly involved in production.
   c. **Manufacturing Overheads:** Costs like utilities, maintenance, and equipment depreciation.

   COGS = Opening Inventory + Raw Material Purchases + Direct Labor + Overheads − Closing Inventory

3. **Gross Profit** is Revenue minus COGS.
   Gross Profit=Revenue−COGS

4. **Operating Expenses**
   Costs are not directly tied to production but necessary to run the business:
   a. Salaries (administrative staff)
   b. Marketing and sales expenses
   c. Rent and utilities for office spaces
   d. Research and development (R&D)

5. **Non-Operating Income and Expenses**
   a. Interest in income or expenses
   b. Gains or losses from asset sales
   c. Other miscellaneous items

6. **Net Profit (or Loss)**
   The final bottom line after accounting for all revenues, costs, and expenses, including taxes.
   Net Profit=Operating Profit + Non-Operating Income/Expenses − Taxes

**The demand and supply curve** in manufacturing is a fundamental concept used to understand how market forces influence production volumes, pricing, and profitability. Here's an overview tailored to manufacturing:

# 1. The Demand Curve in Manufacturing

The demand curve shows the relationship between the price of a product and the quantity demanded by customers. **Downward Sloping**: As the price decreases, the quantity demanded increases (and vice versa). Factors influencing Demand are

a. Price of the Product
b. Customer Preference

c. Substitute Goods
  d. Economic Factors
  e. Market Size

## 2. The Supply Curve in Manufacturing

The supply curve shows the relationship between the price of a product and the quantity producers are willing to supply. **Upward Sloping**: As the price increases, the quantity supplied increases (and vice versa). Influencing factors

  a. Cost of Production
  b. Technology
  c. Input Availability
  d. Government Policies
  e. Market Competition

**3. Equilibrium in Manufacturing** Equilibrium Price and Quantity: Where the demand and supply curves intersect, the market achieves balance.

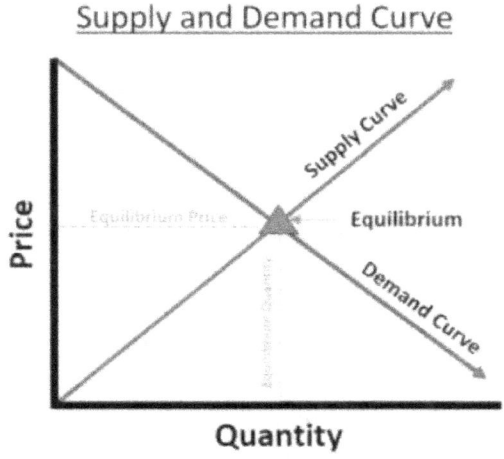

Example of Demand and Supply Curve

# 14. AI in Manufacturing & Factory Floor

Artificial Intelligence (AI) is revolutionizing the manufacturing sector, driving efficiency, precision, and innovation like never before. By leveraging AI technologies, manufacturers can optimize production processes, reduce costs, and stay competitive in an increasingly dynamic market. This book explores the applications, benefits, challenges, and future trends of AI in manufacturing. AI refers to machines and software systems capable of mimicking human intelligence to perform tasks such as decision-making, pattern recognition, and predictive analysis. In manufacturing, AI enhances processes through automation, analytics, and smart systems.

**Machine Learning (ML):** Algorithms that enable systems to learn from data and improve performance over time.

**Computer Vision:** Technology that enables machines to interpret and analyze visual data, such as quality inspection.

**Robotics:** AI-powered robots handle complex tasks with precision and adaptability.

**Natural Language Processing (NLP):** Assists in human-machine communication for process monitoring.

**Predictive Analytics:** Anticipates equipment failures and optimizes production schedules.

Application of AI in Manufacturing

1. Predictive Analytics and Maintenance
2. Quality Control with Computer Vision
3. Supply Chain Optimization
4. Process Automation
5. Design and Prototyping

**Gilbert to AI with Computer Vision Technology**

Computer vision is a branch of AI that enables machines to interpret and analyze visual data, such as images or videos, to make informed decisions. In manufacturing, it is widely used for tasks like defect detection, inventory tracking, and process monitoring. In future I see this technology helping industrial Engineers to do Automated Time study, Defect detections, Spaghetti diagrams and safety, ergonomic analysis all in one package. Real-time inspection systems powered by computer vision can identify defects during production, ensuring only high-quality products move forward. This reduces waste, rework, and customer returns, aligning with lean principles. AI-powered vision systems analyze production workflows, monitor equipment utilization, and identify delays. These insights help manufacturers streamline operations and eliminate non-value-adding activities. monitoring equipment in real time, computer vision can detect wear and anomalies, predicting potential failures. This supports proactive maintenance, keeping production lines running smoothly. AI-powered systems can track inventory levels visually, automatically updating stock data and ensuring alignment with production demands. AI-driven vision systems can monitor workspaces for safety hazards, such as improper equipment handling or potential ergonomic risks, promoting a safer, more efficient environment.

**Future of Computer Vision and Lean Manufacturing**

The synergy between computer vision, lean manufacturing, and AI is set to grow as technology advances. Future developments may include:

- **Collaborative Robots (Cobots):** Equipped with vision systems to work alongside humans efficiently.
- **Smart Factories:** Fully automated environments were computer vision and AI drive end-to-end lean operations.

- **Advanced Analytics:** Deeper insights from computer vision data for strategic decision-making.

As an Industrial Engineer aiming to integrate Artificial Intelligence (AI) into your work, acquiring a mix of technical, analytical and domain specific skills is essential.

1. Programming and coding with Python, R METLAB
2. Machine Learning (ML) Deep Learning and Nural networks
3. Data Science and Analytics
4. Optimization Techniques- Linear programming, Dynamic programming, Genetic algorithms
5. Computer Vision
6. Robotics and Automation
7. Simulation and Modeling
8.

As we come to the end of this journey, it's clear that the future of industrial engineering is defined by innovation, adaptability, and a relentless drive for excellence. The integration of AI, automation, and data-driven decision-making is transforming the way we design, manage, and optimize manufacturing systems. But more importantly, it's empowering us to redefine what's possible.

As an industrial engineer, you have the power to shape this future, to lead change, and to push the boundaries of what can be achieved. The tools and strategies you've learned here are not just concepts, they are pathways to becoming part of the top 1% in your field. By embracing the potential of AI and continuously refining your skills, you are positioning yourself at the forefront of a revolution that will continue to evolve.

The journey doesn't end here. Every project, every challenge, every breakthrough is an opportunity to grow, to innovate, and to make an impact. As you move forward, remember that the most successful engineers are those who are always curious, always learning, and always striving to improve the world around them.

Thank you for joining me on this exploration of the future of industrial engineering. I have no doubt that you will take these insights and apply them to achieve extraordinary things. The road ahead is full of possibilities—take the first step today and lead with confidence.

**The best is yet to come.**

# References

1. The Goal: 40th Anniversary Edition: A Process of Ongoing Improvement by Eliyahu M Goldratt
2. Introduction to Industrial Engineering by Scott Mackay
3. The Complete Guide to Mix Model Line Design by Gerald Leone & Richard D Rahn
4. Industrial Engineering and Production Management by Martand Telsang
5. Introduction to Industrial and System Engineering by Robert Wayne Atkins
6. Assembly Line Design by WE-MIN CHOW
7. Assembly Line Design by Brahim Rekiek & Alain Delchambre
8. Work Systems by Mikell P. Groover
9. Wikipedia.com
10. Reserchgate.net
11. https://blog.kainexus.com/improvement-disciplines/kaizen/kaizen-event/kaizen-event-planning-in-7-simple-steps
12. https://weeverapps.com/5s/5s-overview/
13. https://global.toyota/en/company/vision-and-philosophy/production-system/
14. https://www.lean.org/lexicon-terms/operator-balance-chart/
15. https://www.vaia.com/en-us/explanations/microeconomics/supply-and-demand/

www.ingramcontent.com/pod-product-compliance
Lightning Source LLC
Chambersburg PA
CBHW071419220526
45469CB00004B/1349